Main Battle Tanks Coloring Book For Teens & Adults

Military & Army - Grayscale Line Art Colouring - Abrams, Merkava, Challenger, T-72, WW2

Rachel Mintz

Copyright © 2017 Palm Tree Publishing - All rights reserved. No part of this publication may be reproduced, distributed, or transmitted in any form or by any means, including photocopying, recording, or other electronic or mechanical methods, without the prior written permission of the publisher, except in the case of brief quotations embodied in critical reviews and certain other noncommercial uses permitted by copyright law.

"If the tanks succeed, then victory follows."
— Heinz Guderian

M10 (USA)

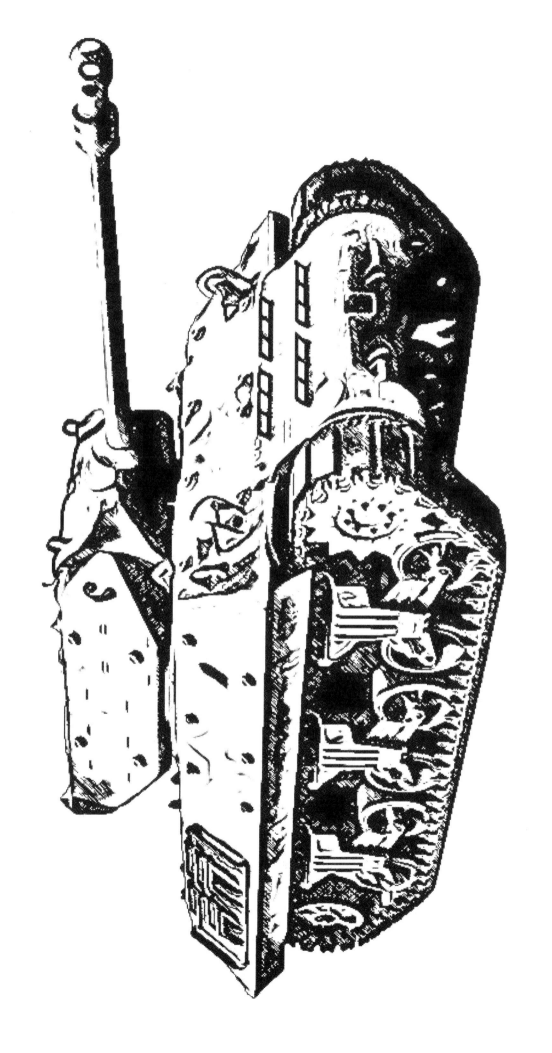

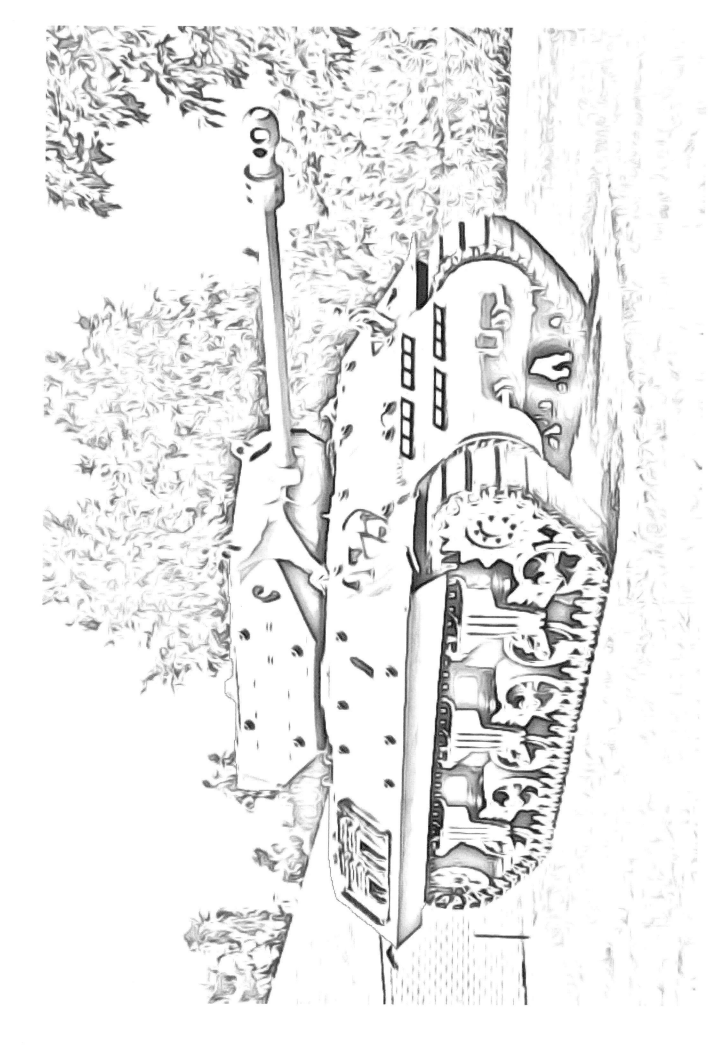

Centurion Mark 5 (UK)

Abrams M1A2 (USA)

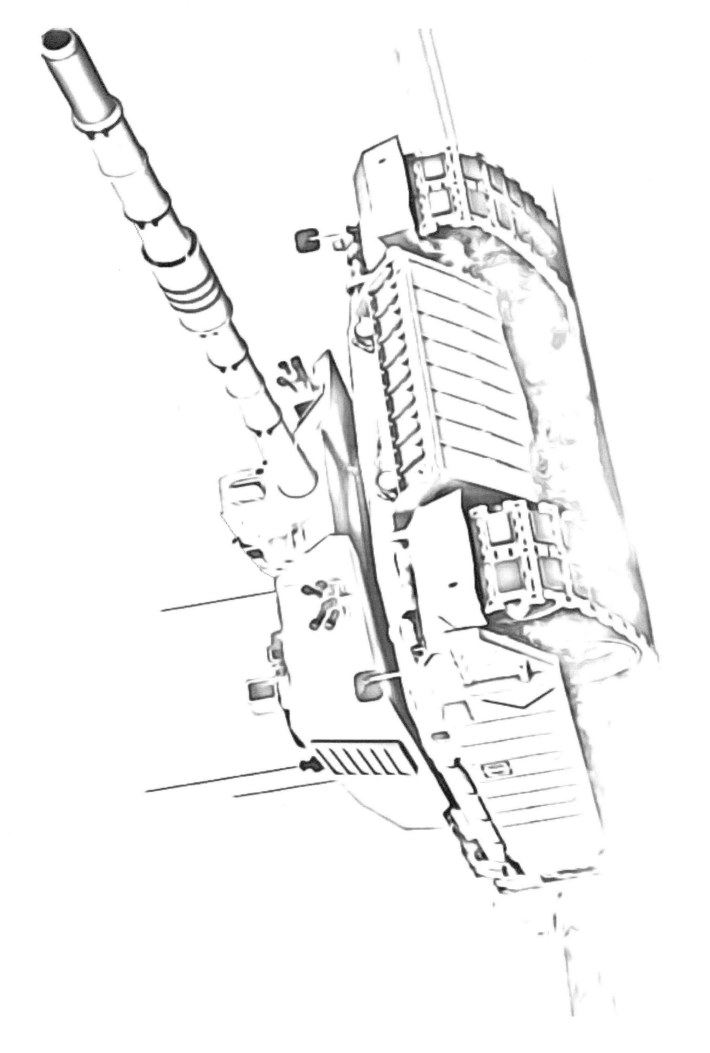

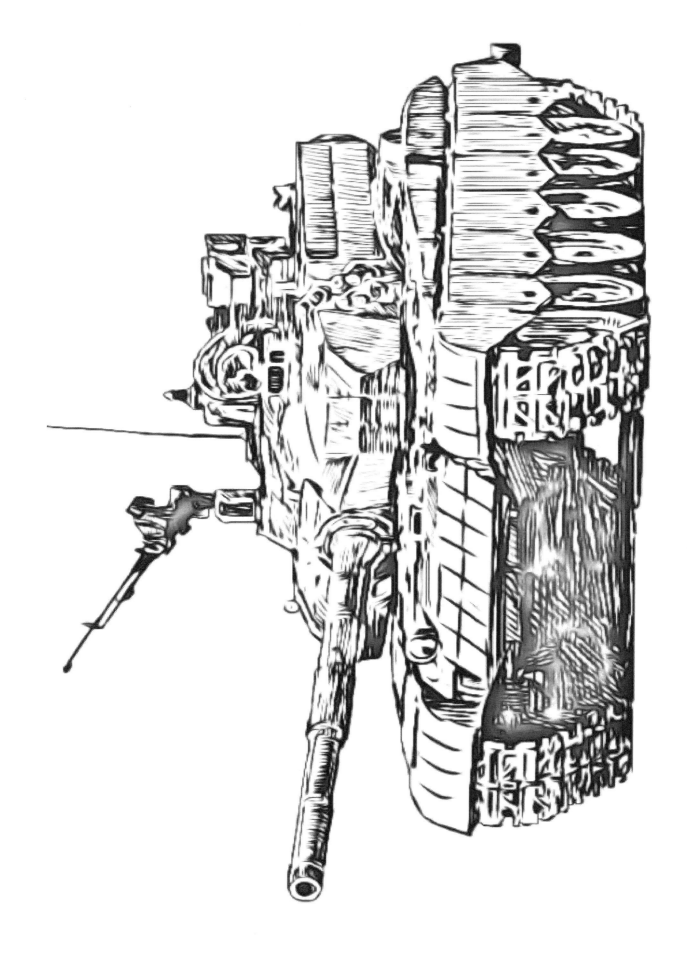

T-55 (Russia)

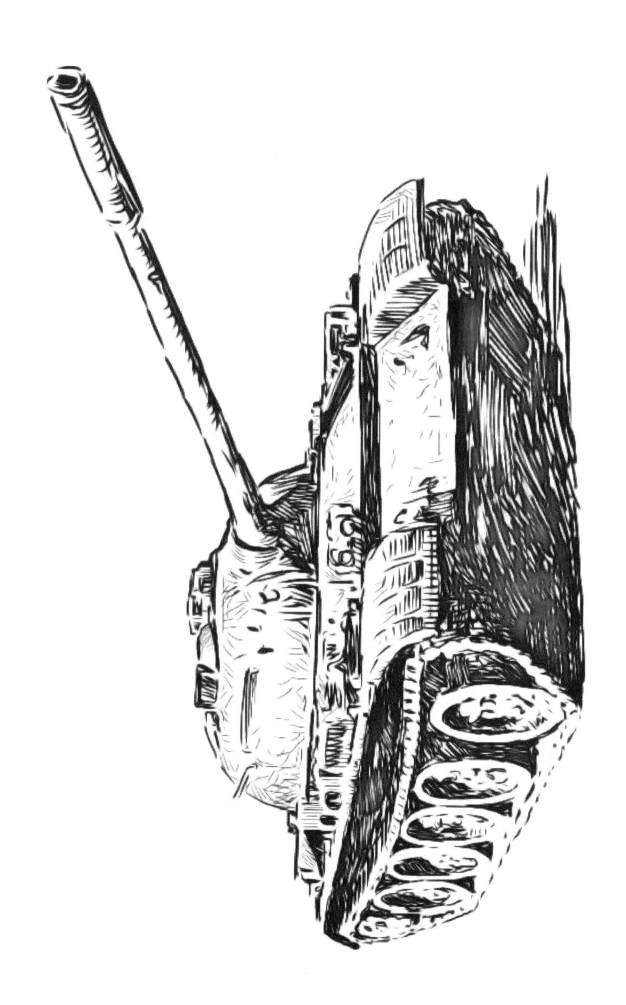

Merkava Mark 4 (Israel)

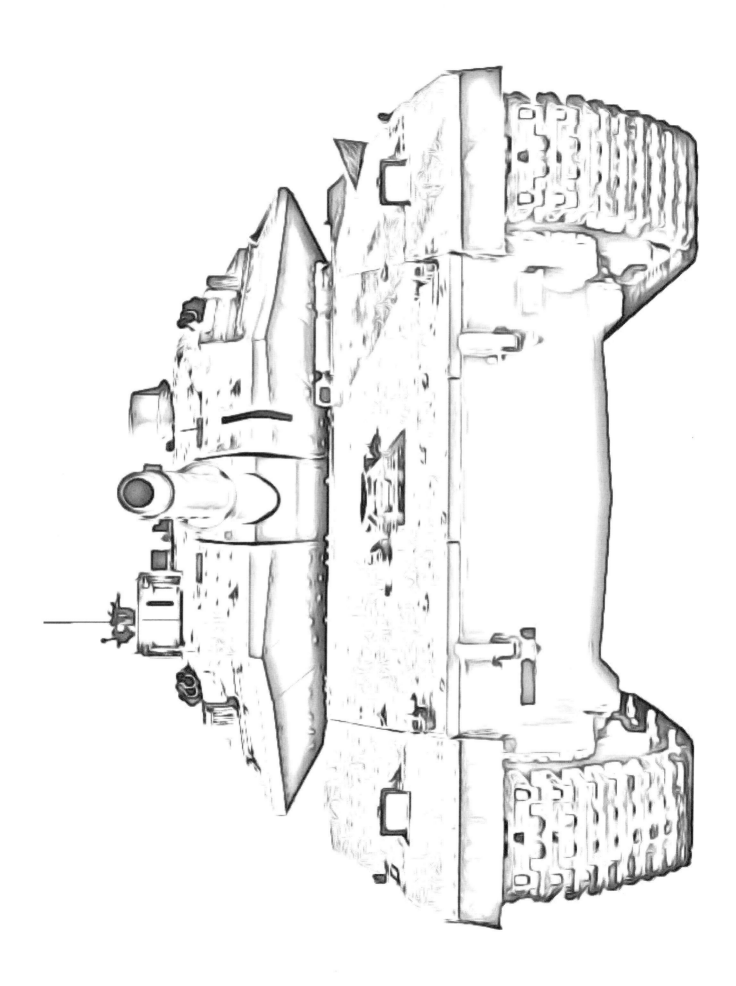

Merkava Mark 4 (Israel)

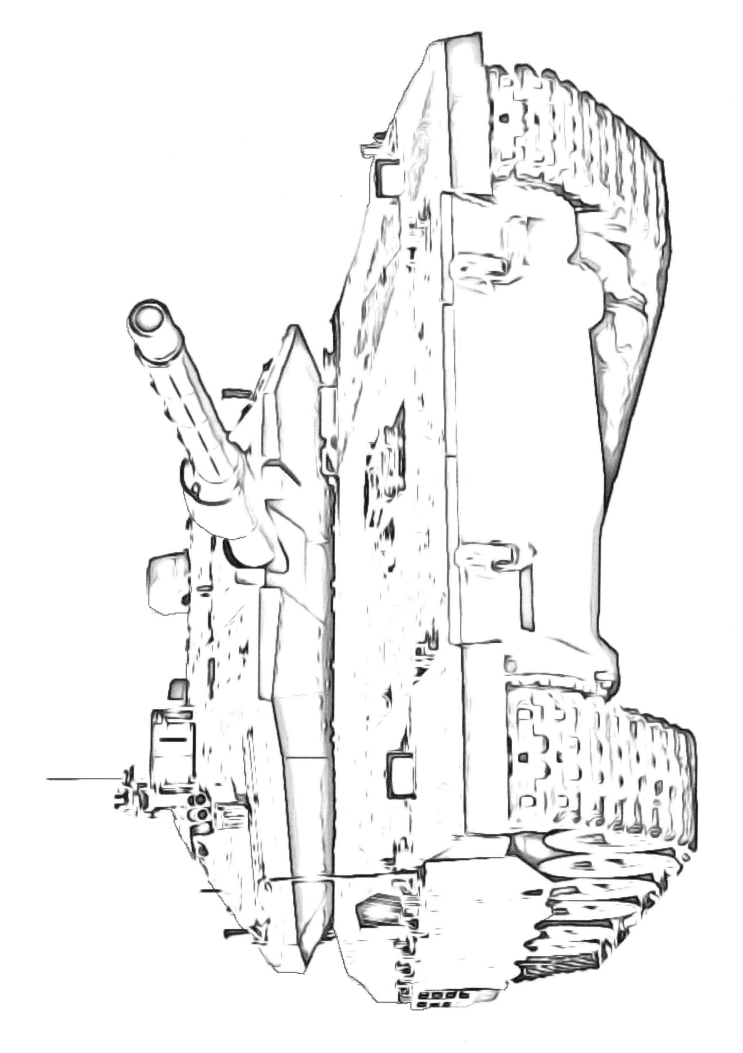

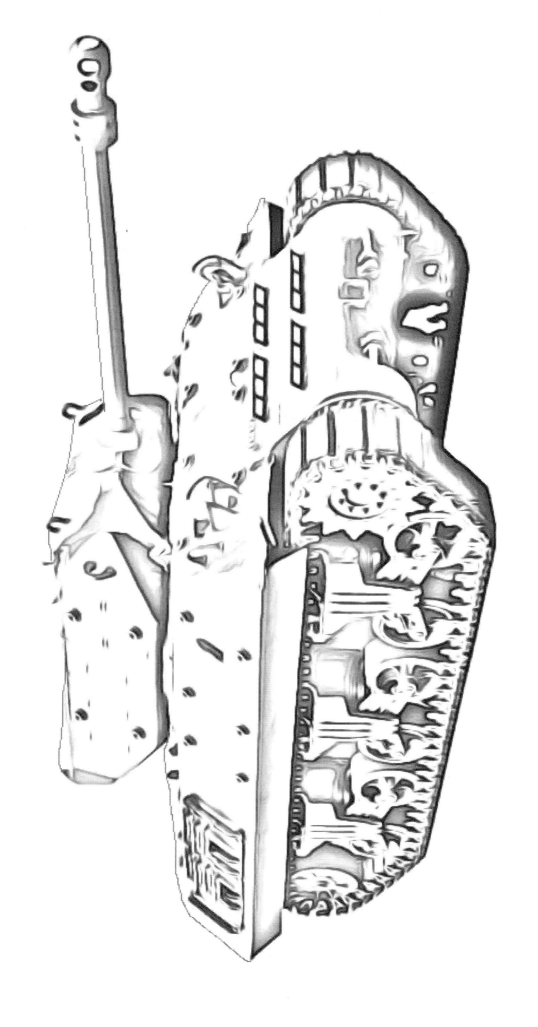

BMP 2 (USSR)

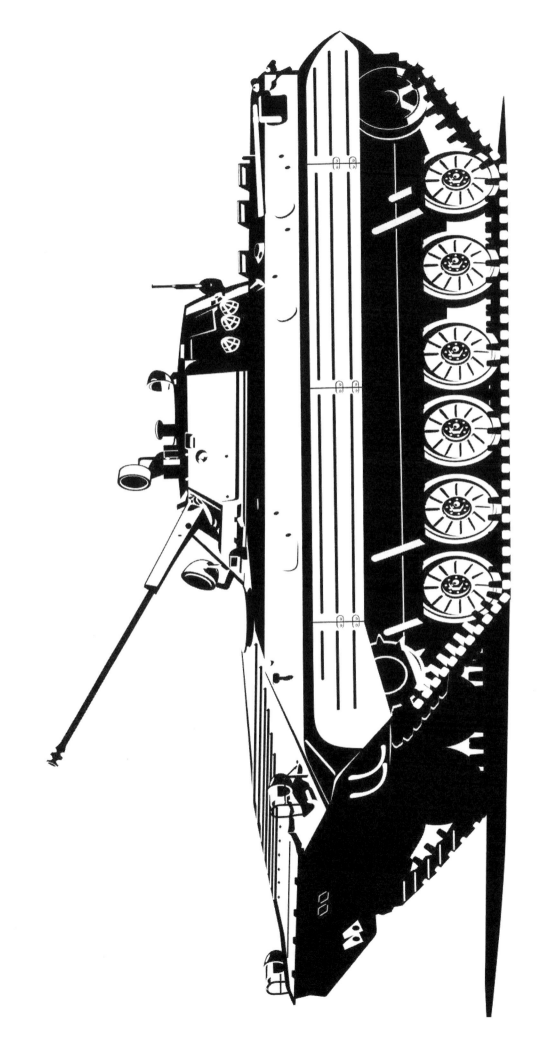

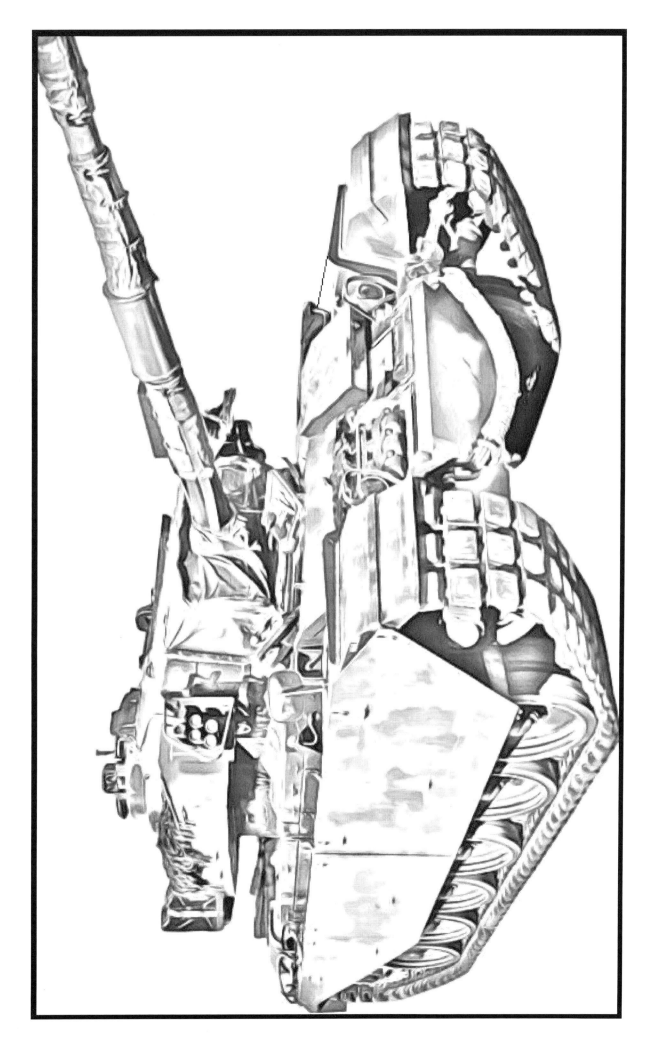

Type 99 (China)

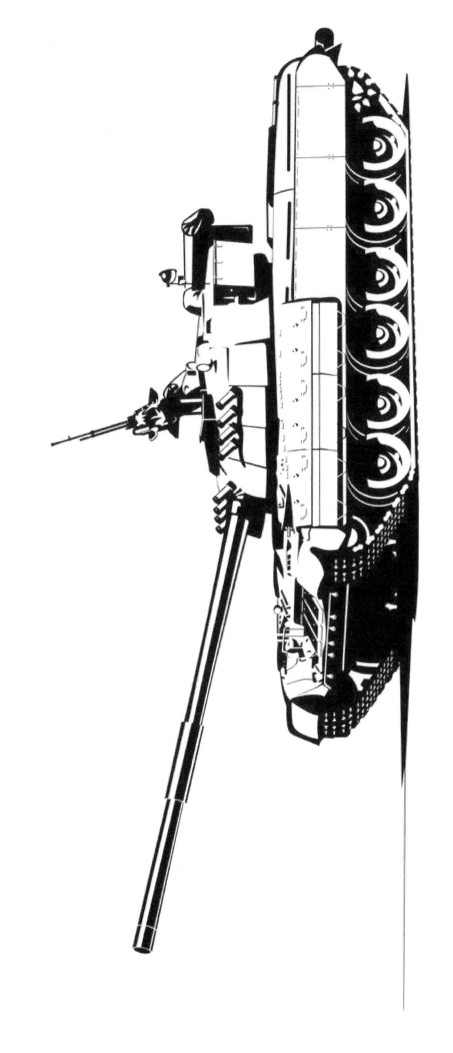

Abrams A1M1 (USA)

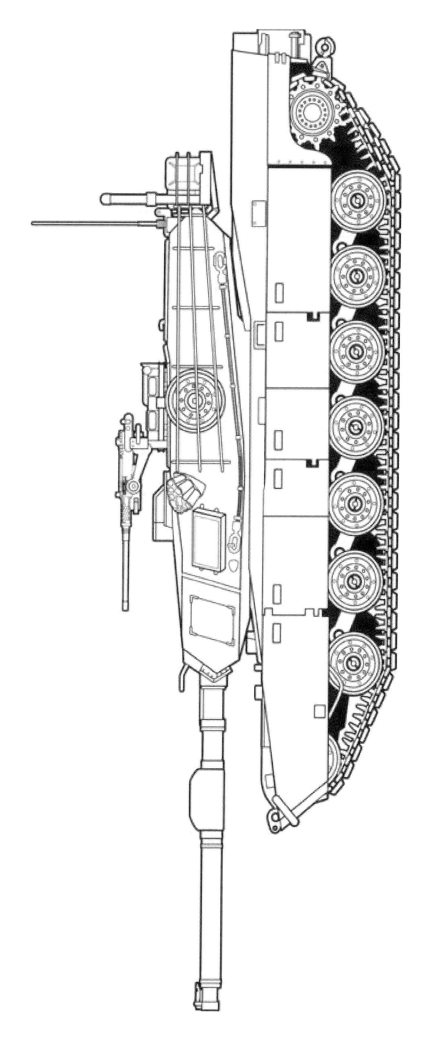

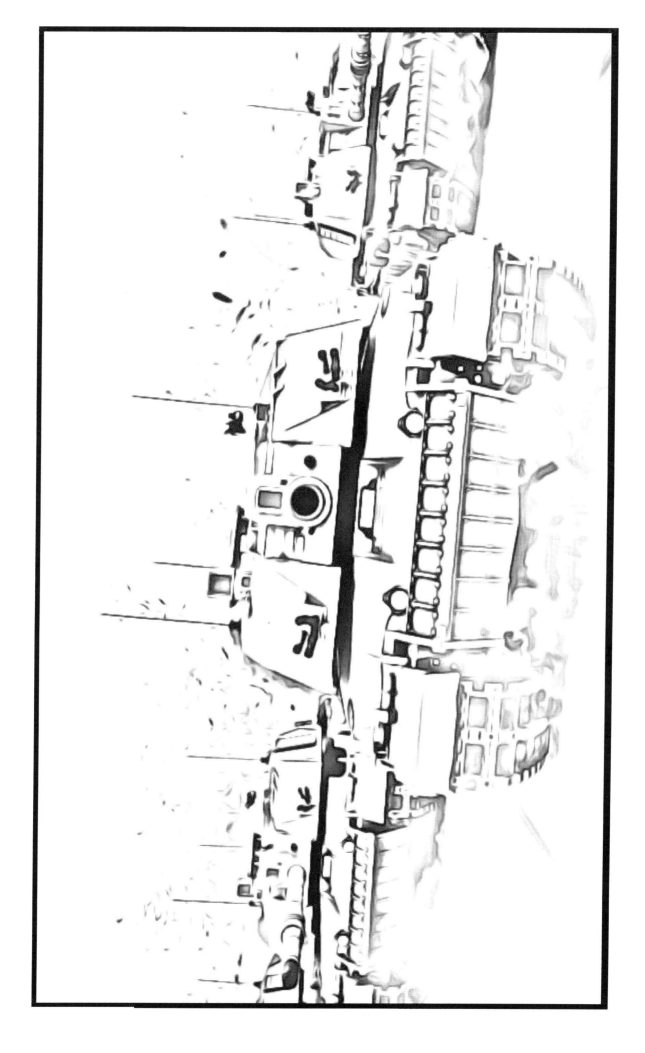

AMX 13 (France)

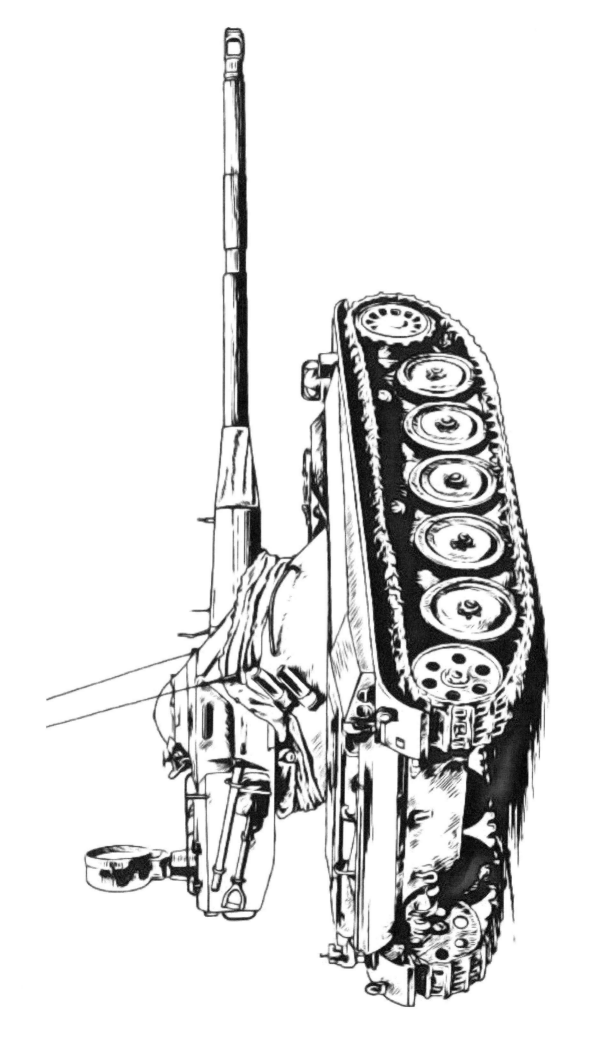

T-34 (USSR)

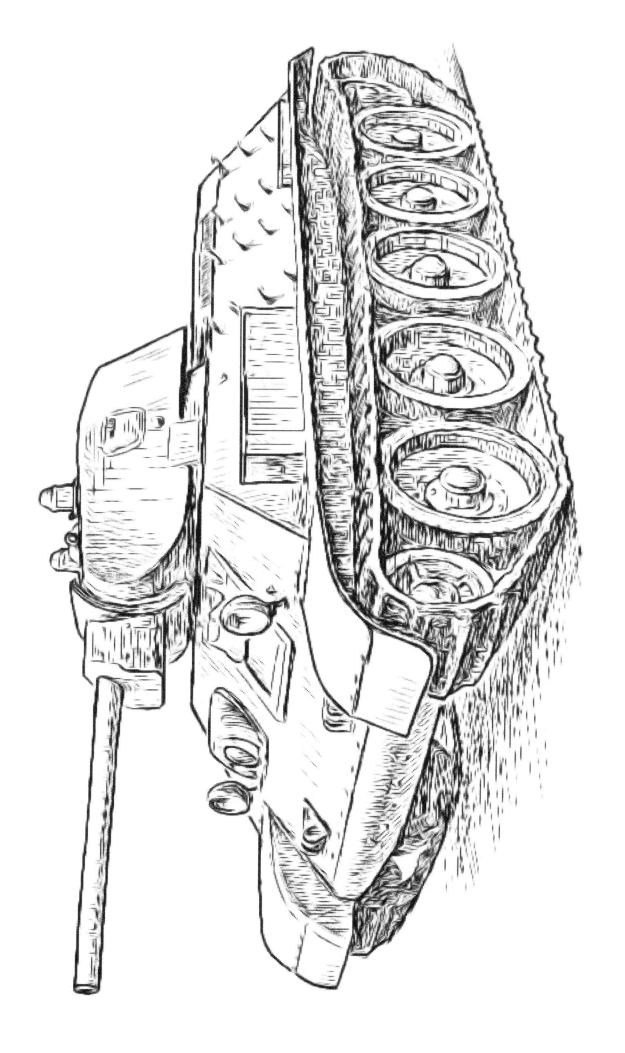

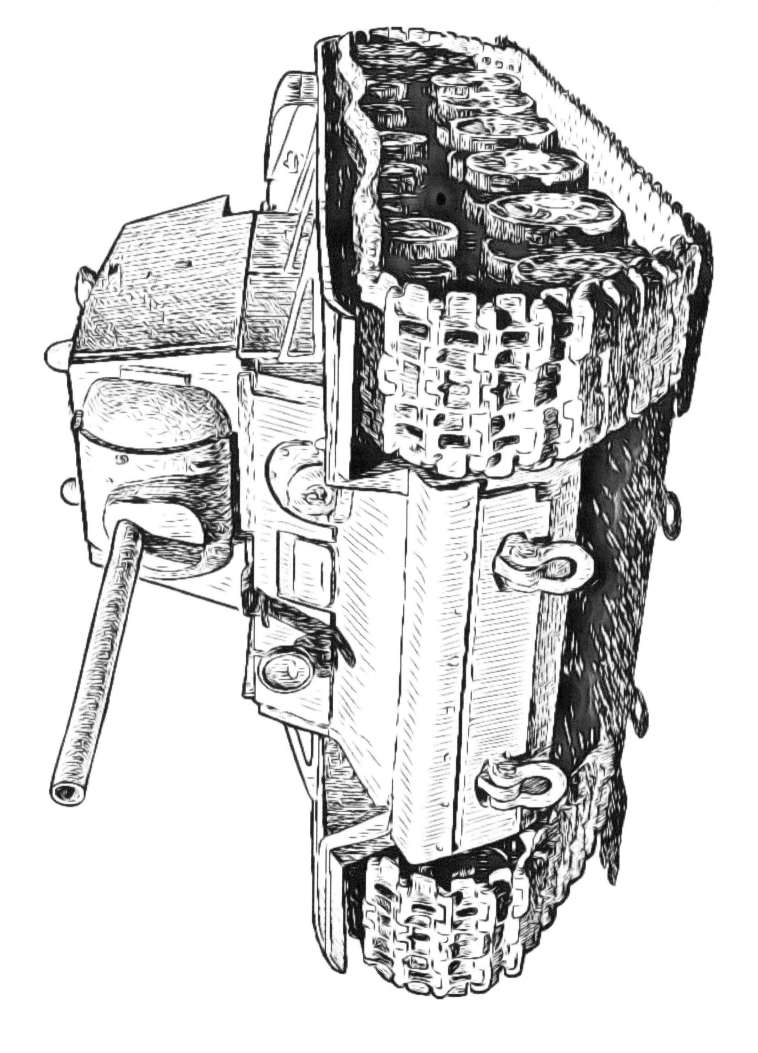

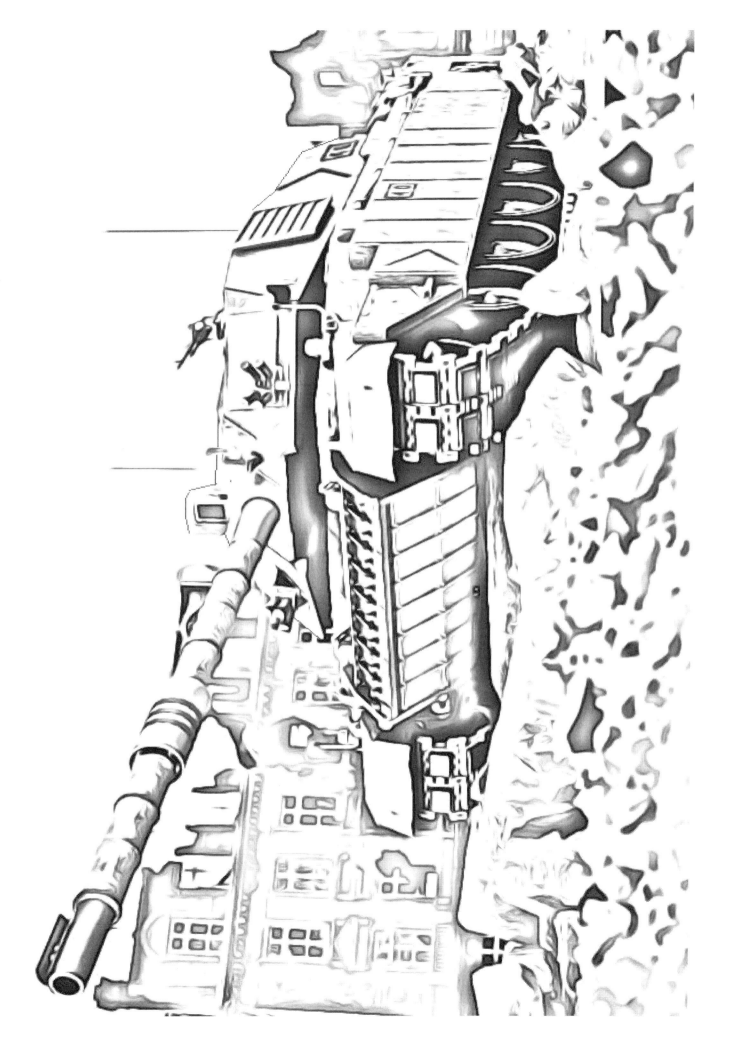

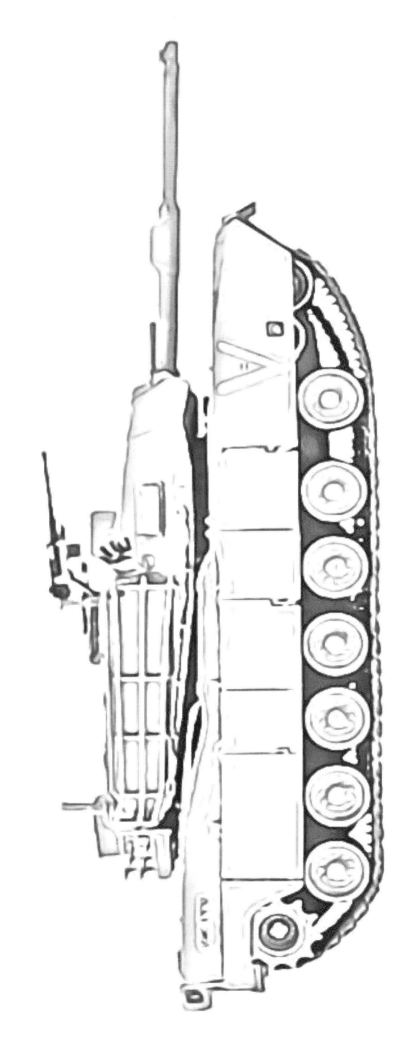

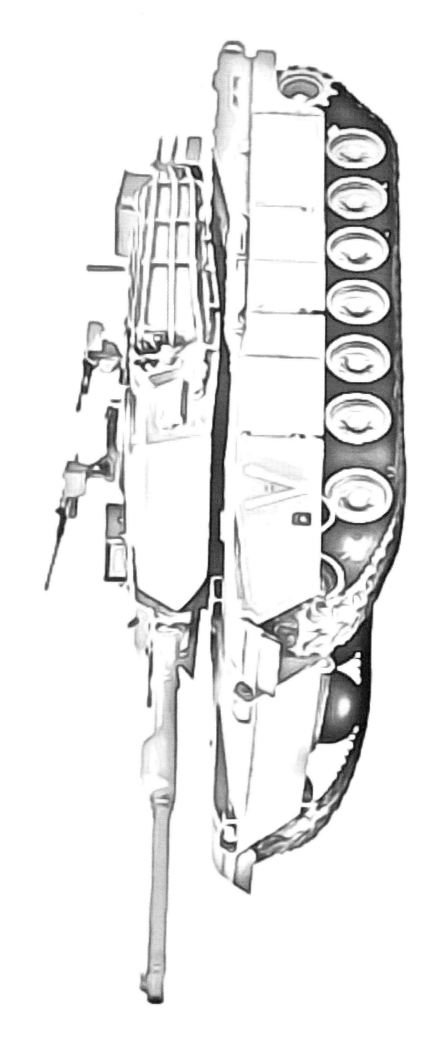

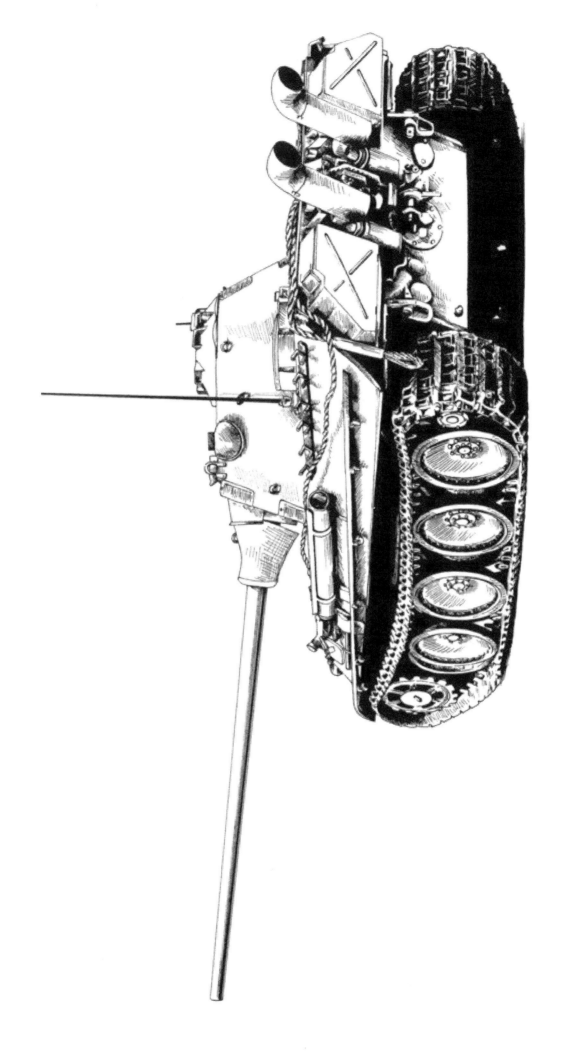

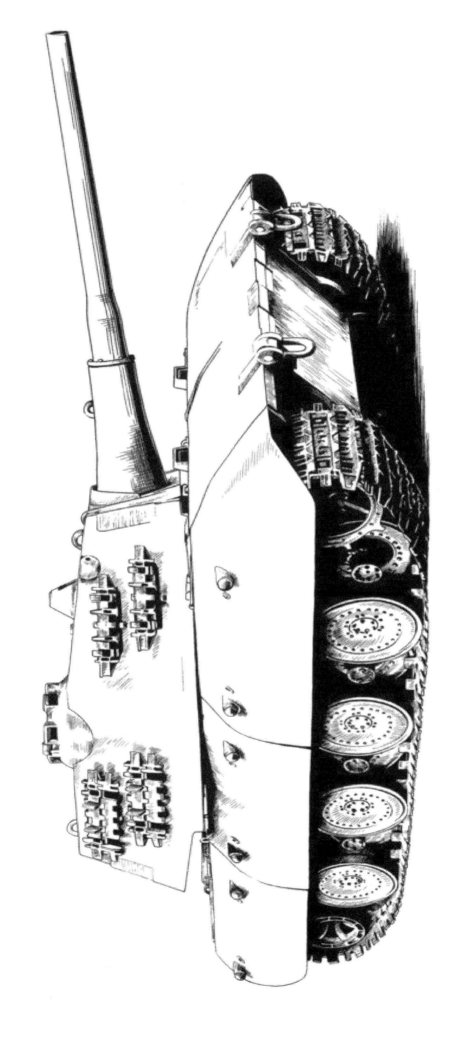

Challenger 2 (UK)

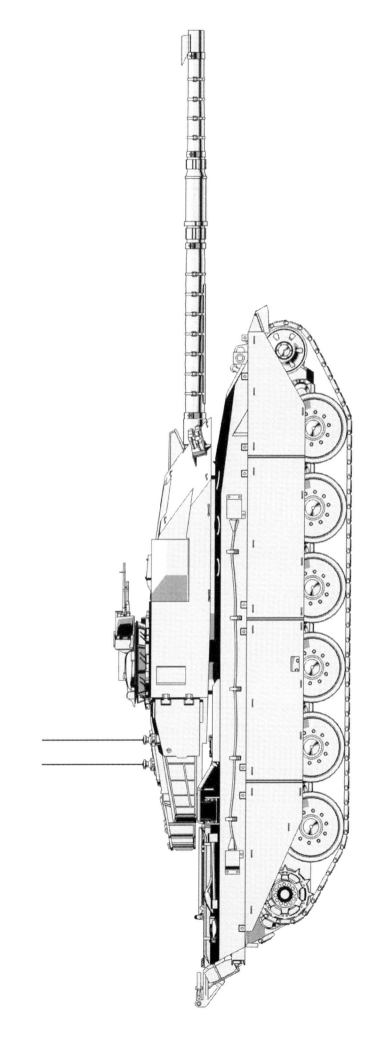

Challenger 2 (UK)

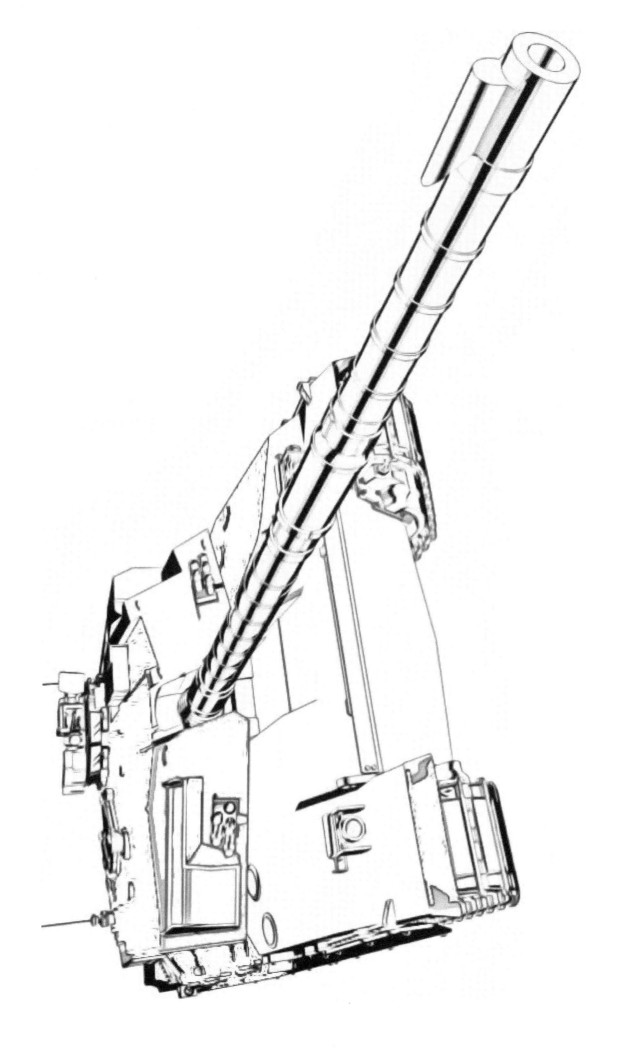

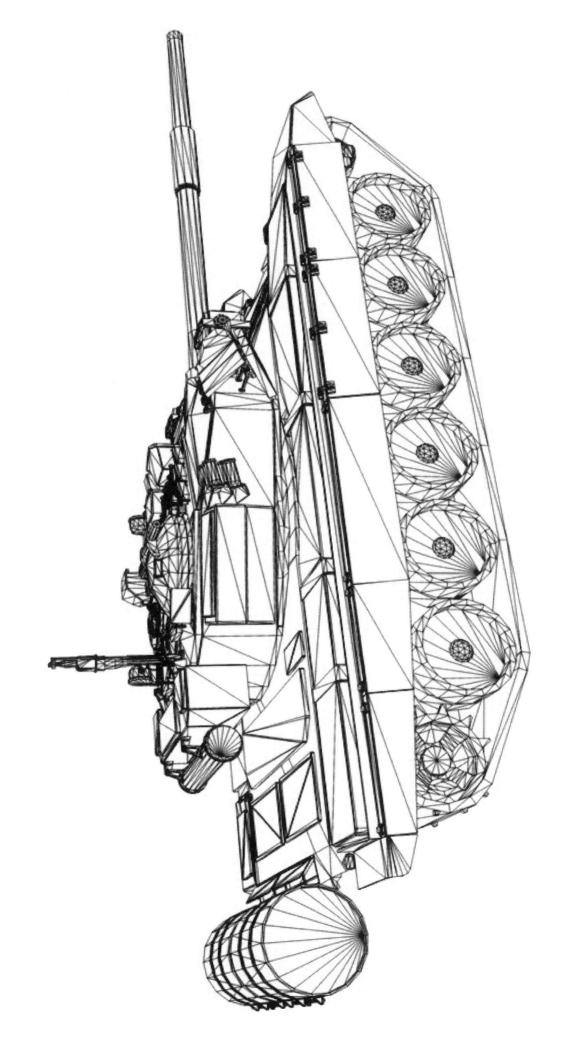

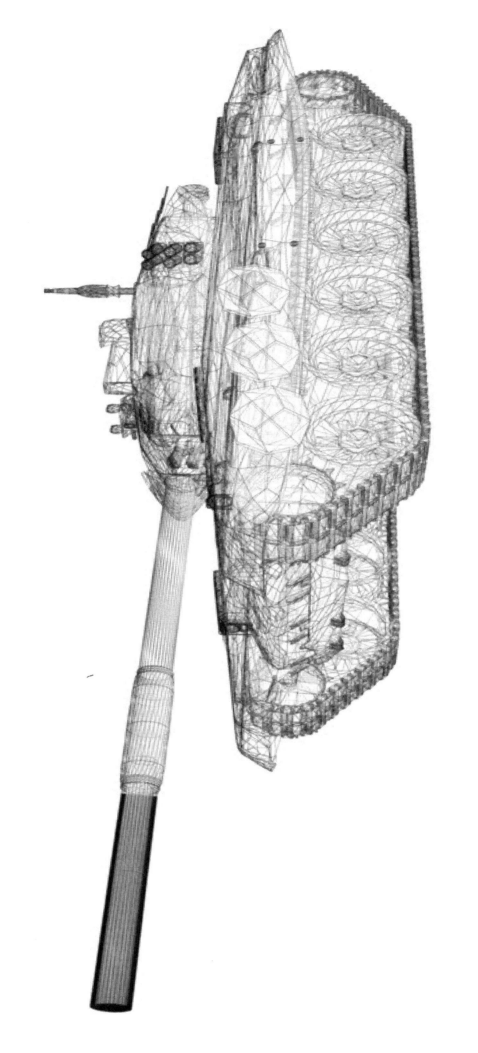

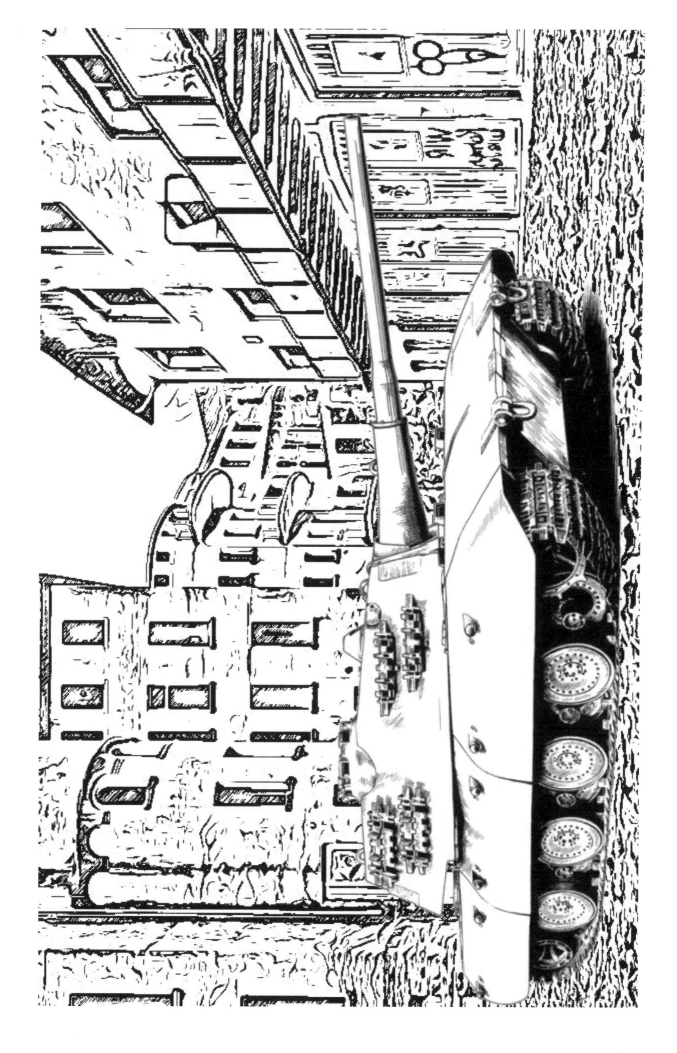

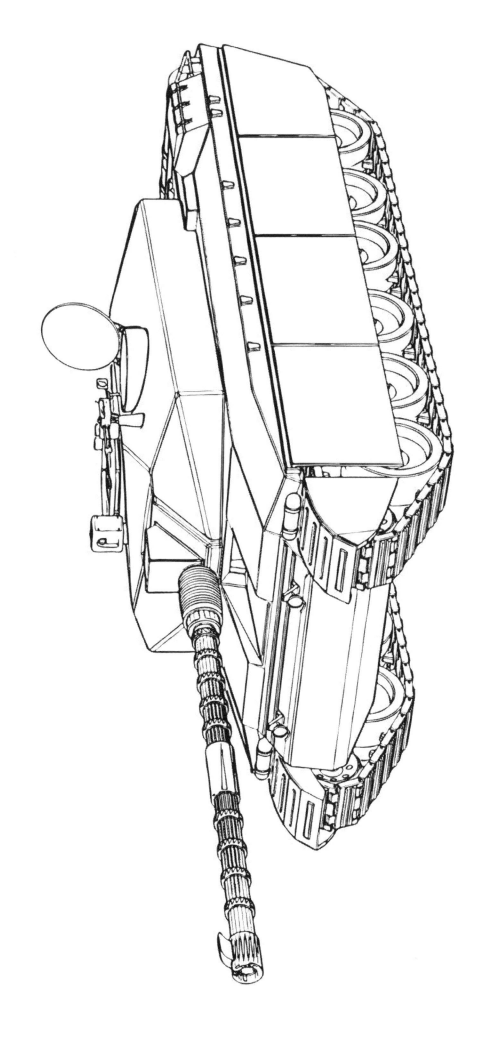

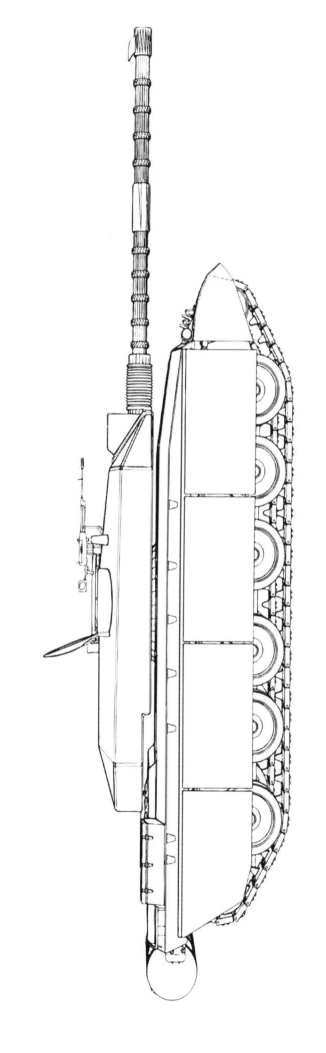

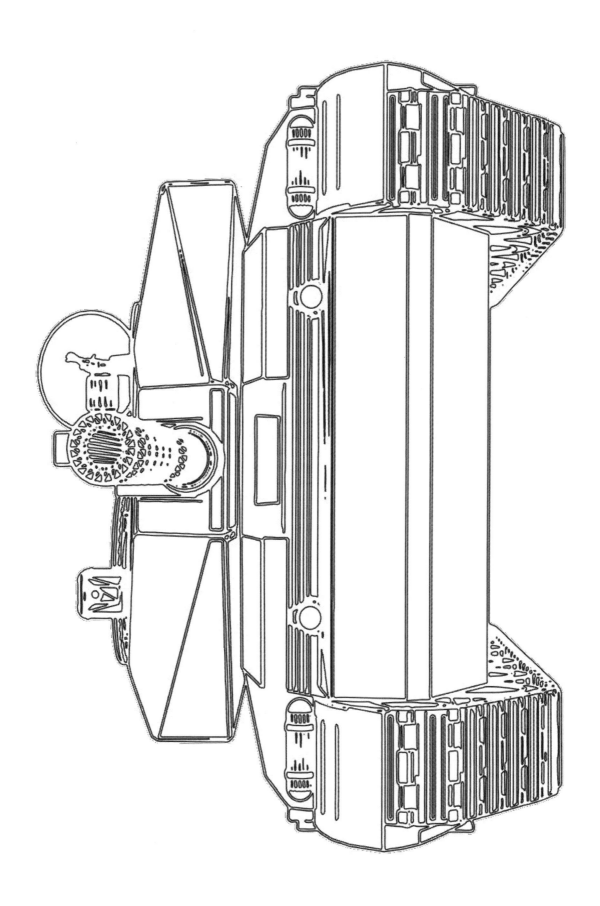

T-72 (USSR)

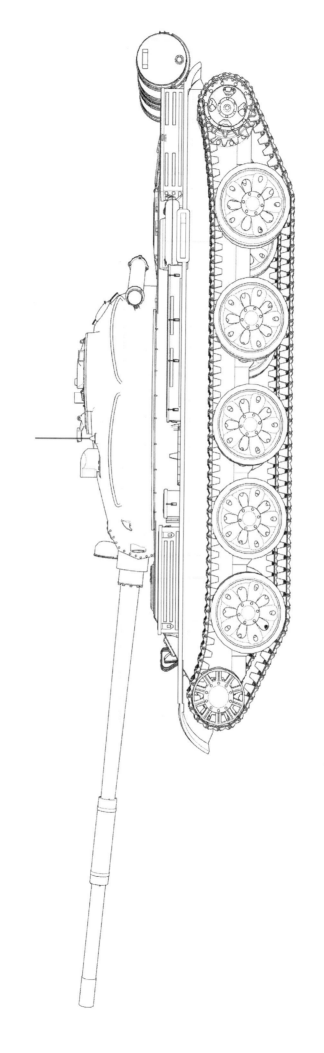

T-72 (USSR)

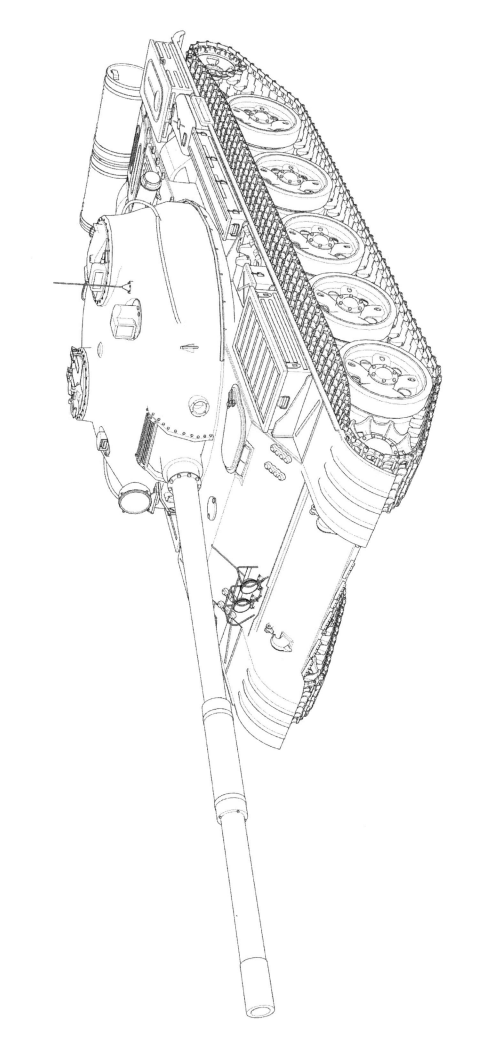

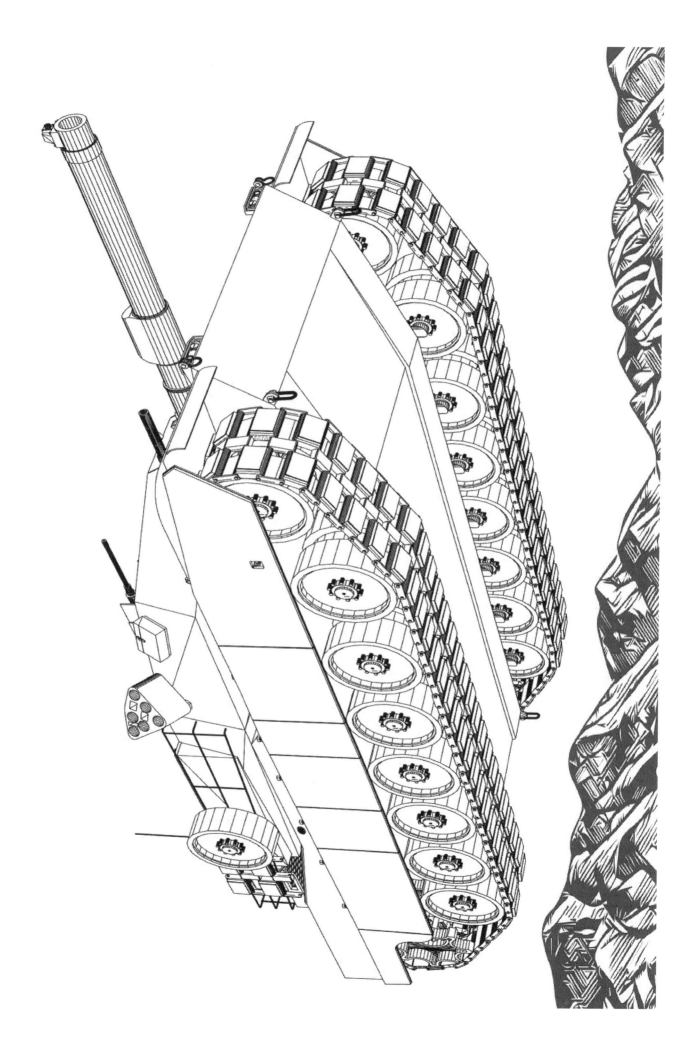

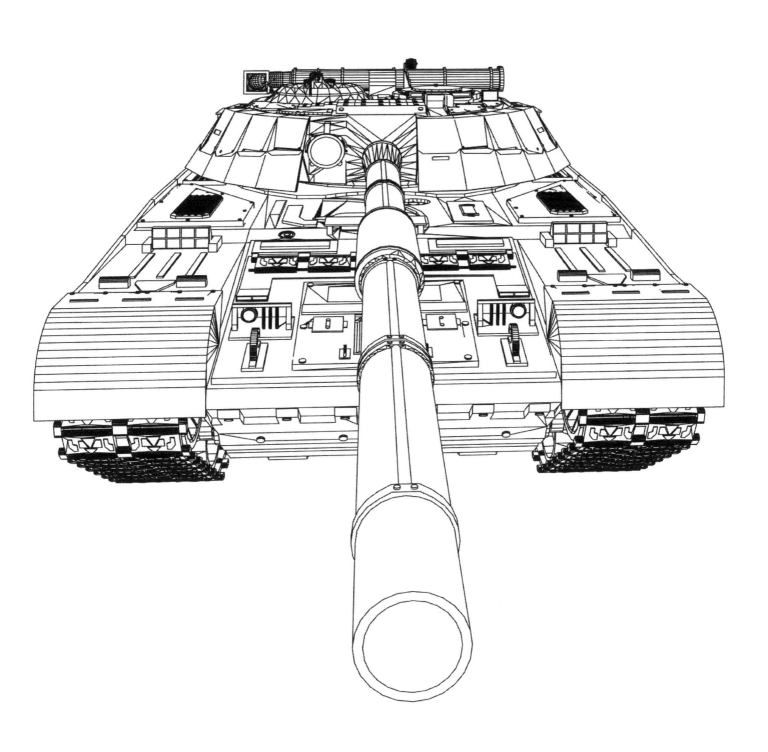

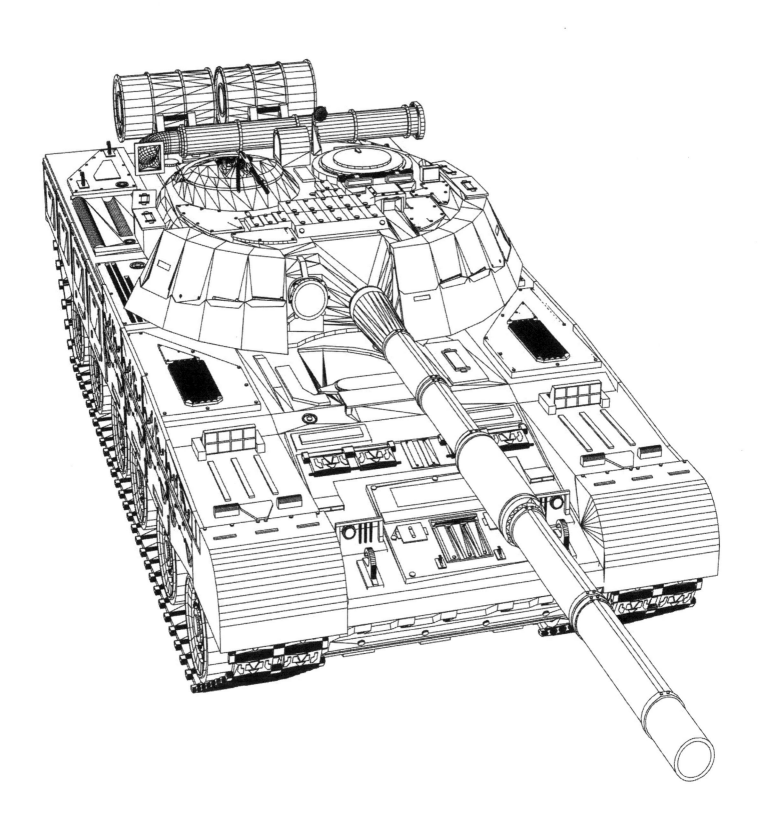

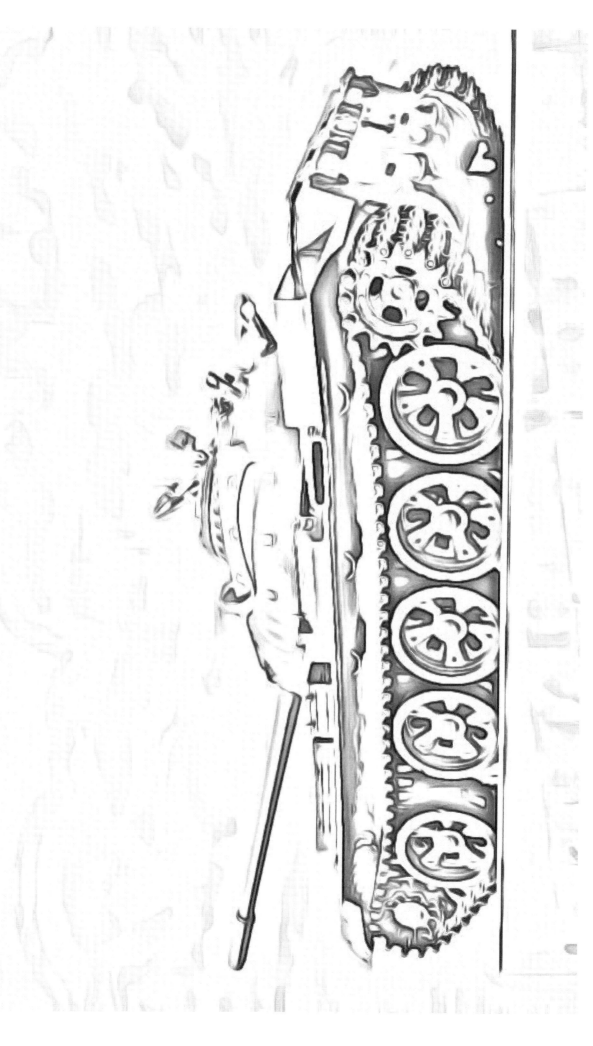

How to spot the difference between USSR tanks and Western tanks?

There are several differences, them main ones are:
The type of suspension used,
Shape of turret
Location of fume extractor.

Tracks suspension:
Russian tanks T34,T-55, T-62, T-72, have the upper part of the tracks laying on the road wheels as can be seen in the T55 below. T-72 and T-90 have their tracks slightly lifted, but still lower than the tracks of western tanks.

Western tanks have the upper tracks tension lifted above the road wheels. As seen in the lower image.

Another difference is the size and shape of the turret.

Russian tanks have 3 crew members so they have smaller and rounded turret. Smaller turret means smaller silhouette in battlefield. Having 3 crew members only is possible because they have an auto-loading mechanism for the main gun rounds. See the T-55 on the next page (top). Western tanks have 4 crew members and have larger, wider and longer shaped turrets. See the Abrams and Merkava on the bottom of the next page.

The fume extractor location can also distinguish Russian tanks.

In most of the Russian tanks the fume extractor is located closer to the end of the main gun barrel than to the turret. This helps gain higher velocity for the rounds, but also means more fumes for the crew inside turret after each round is shot.
In the T-55 it is at the very end (top of next page). The T-72 has it on the front third part (second tank on next page).

Western tanks have the fume extractor closer to the turret as can be seen in the Abrams and Merkava at the bottom of the next page.

Why does the Merkava look so different than other tanks?

The Israeli Merkava tank is quite a unique tank, which does not resemble other Western tanks. The Merkava has similar performance to Western main battle tanks, but has some unique features.

The tank was designed after the heavy Yom Kippur War. In that war hundreds of Israeli tanks were hit by Sager ATGMs, and hundreds of crew members were injured and killed. 90% of those hits were from the front and 45 degrees to the sides. So when the Israeli designers sat down to design the next generation tank, their main focus was to plan a tank that will serve maximum crew safety!

The Merkava designers placed the engine upfront, making the front part of the tank heavily protected like no other tank in the world.
On the next page you can see the location of the engine in the Russian T-72, British Challenger, USA Abrams and Israeli Merkava.

Having the engine at the front, the Merkava rear hull became available for carrying troops. The rear hull allows crew or troops to enter/leave the tank from the rear in a protected way. It takes little adjustments, and the Merkava can also carry up to 8 infantry soldiers **inside** it.

The Merkava turret has a futuristic sloped armor shape, lowering the penetration angle of incoming shells and missiles to minimum. The armor is placed over the turret in modules like 'Lego' parts, so when the tank is hit and a module is destroyed, that module can be easily replaced even in the battlefield and the tank downtime is minimum.

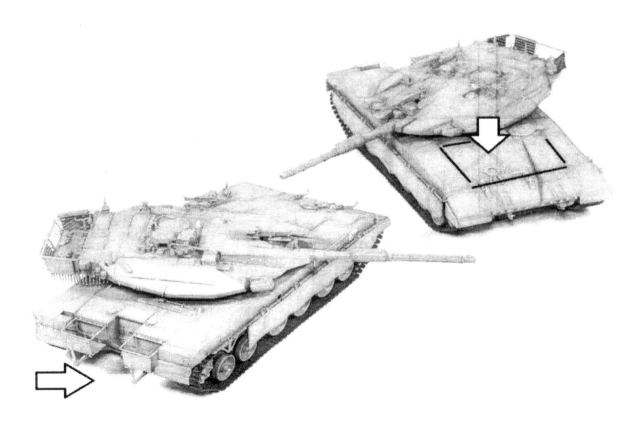

Engine location in main battle tanks vs. Merkava:

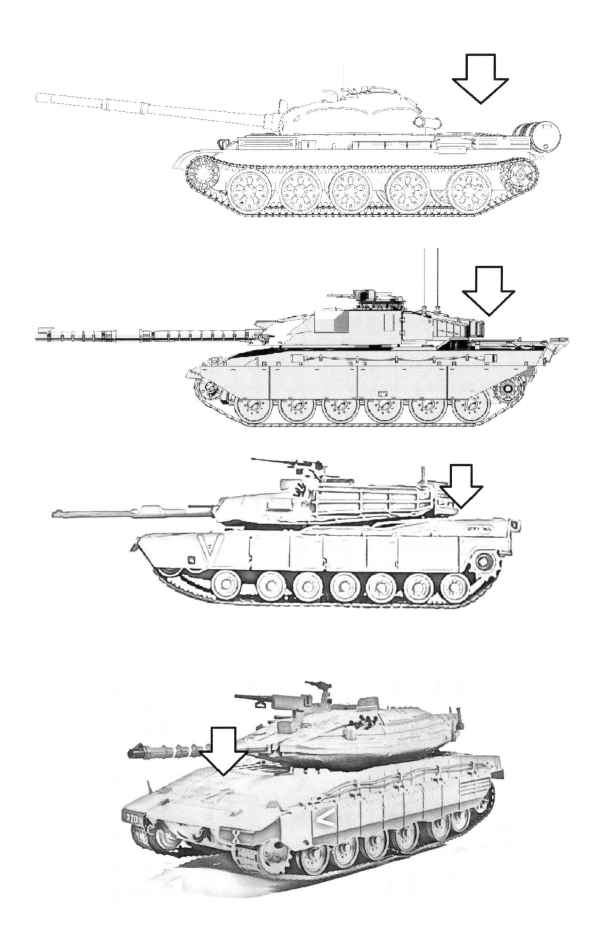

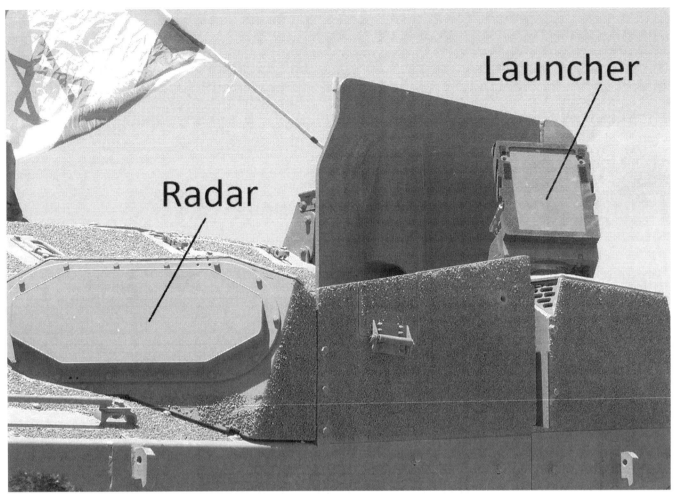

By NatanFlayer [CC BY-SA 3.0 (https://creativecommons.org/licenses/by-sa/3.0)], from Wikimedia Commons

The **Trophy Trophy Active Protection system** developed by Israel, has managed intercept dozens of anti-tank weapons including Kornet, Metis and RPG-29 in intense close range combat situations.

This is another step in the tank and Anti-tank missile race.. Giving the Merkava an edge over incoming missiles and RPGs.

Thank you for coloring with us!

We hope you had a good time and found the information interesting. Let us know by adding your review at Amazon. We are thankful for every new review.

More coloring books you can order:

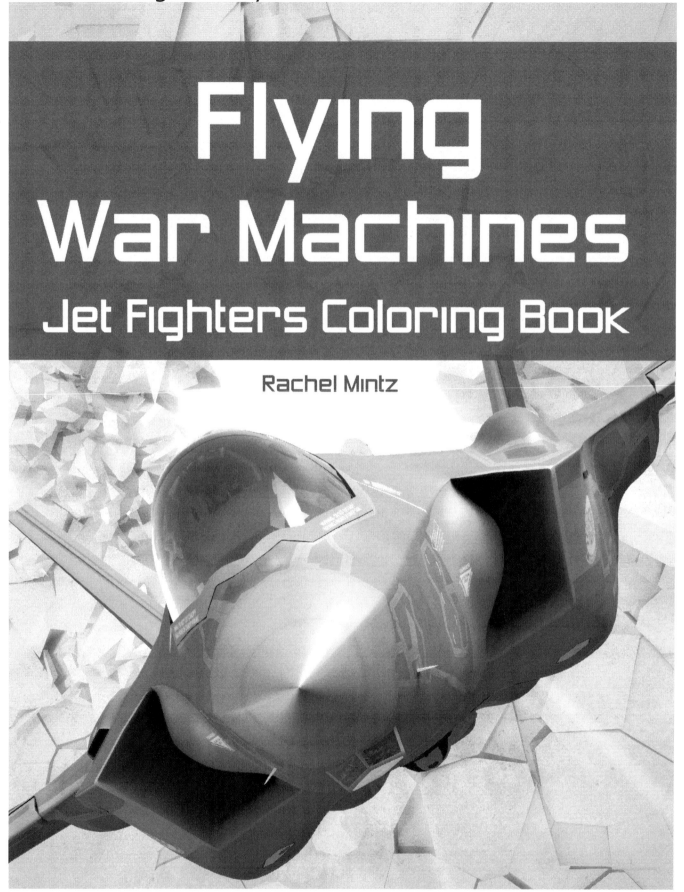

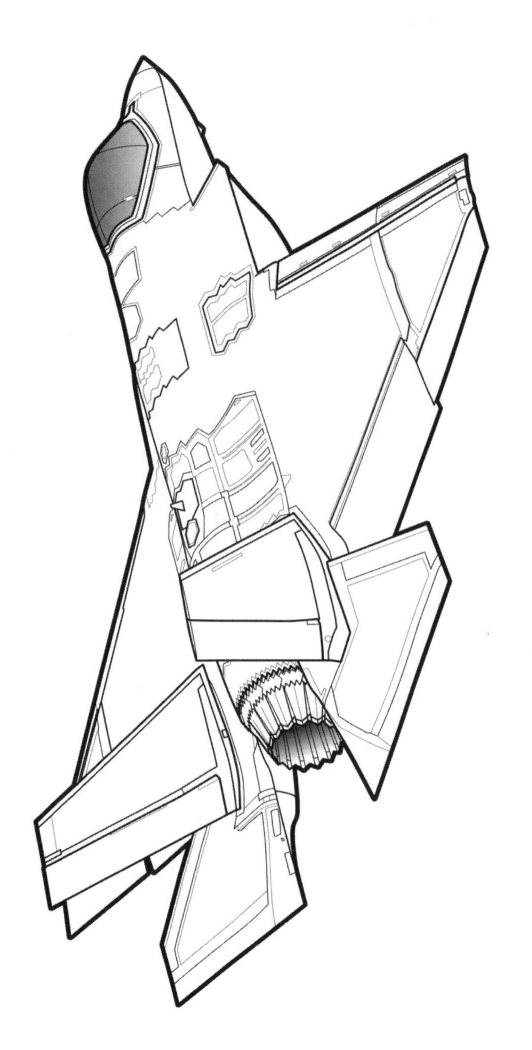

Lethal War Machines
Military Coloring Book

Rachel Mintz

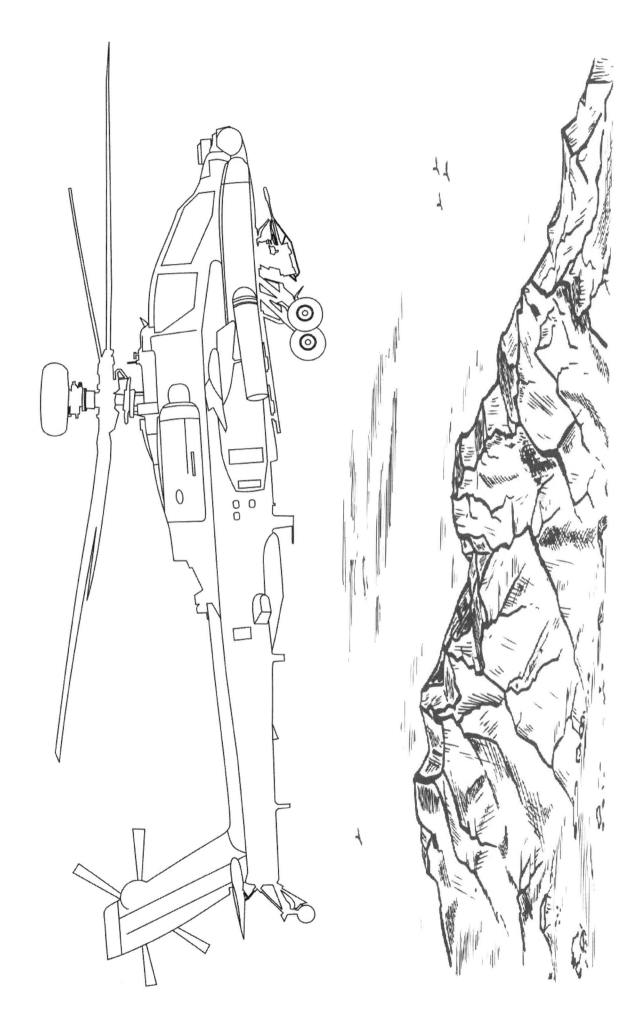

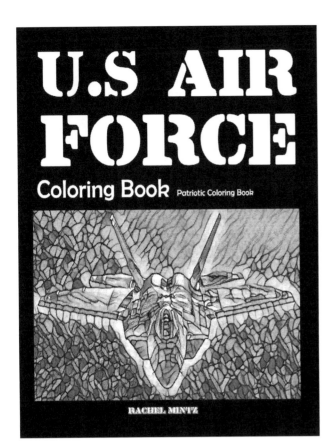
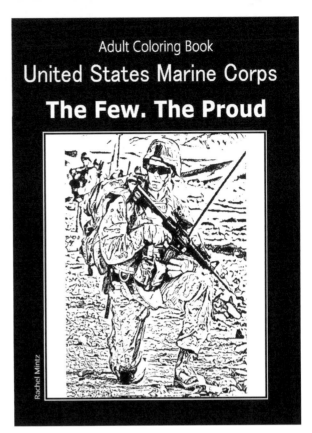
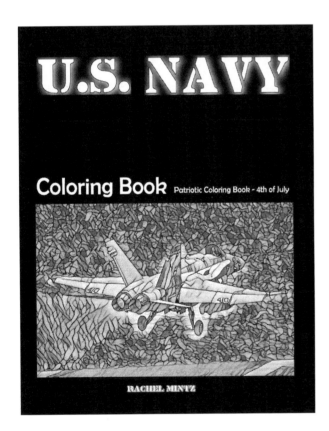
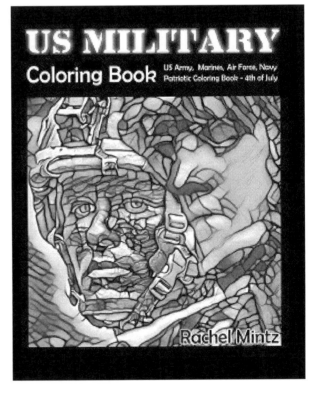

SPECIAL FORCES

COLORING BOOK FOR ADULTS

RACHEL MINTZ

Thank you for coloring with us

Made in the USA
Lexington, KY
22 October 2018